I0070856

Radio Propagation Laboratory

Stanford Electronics Laboratories

Stanford University

Stanford, California

THEORY OF RADAR STUDIES OF THE CISLUNAR MEDIUM

by

V. R. Eshleman

R. C. Barthle

P. B. Gallagher

Scientific Report No. 5

July 8, 1959

Prepared under Air Force Contract AF19(604)-2193.

The research reported in this document has been sponsored by the
Electronics Research Directorate of the Air Force Cambridge Research
Center, Air Research and Development Command.   The publication of
this report does not necessarily constitute approval by the Air
Force of the findings or conclusions contained herein.

SUMMARY

The ionized medium between the earth and the moon is being studied at Stanford University by means of lunar radar echoes. Of prime interest is the ion density beyond the earth's ionosphere, in regions (at distances of several to 60 earth radii) where the interplanetary gas may be dominant. There are several techniques based upon h-f (3-30 Mc) moon echoes for measuring the total integrated ion density. The ionospheric part may be determined separately and subtracted from the total.

The moon's radar range exceeds the true range, especially at low frequencies, because of group retardation, which is a measure of integrated ion density. (The retardation varies as f^{-2}, being 250 μsec at 20 Mc for an average cislunar density of $10^9/m^3$.) Range measurements giving density would require high peak power (for pulse ranging) or high equipment stability (for c-w ranging); furthermore, the true lunar range is not known accurately.

In two of the methods described here, the full average-power capability of an h-f transmitter can be used without regard to high stability or precise lunar range. In one method, three equally-spaced frequencies are sent in the form of an amplitude-modulated wave. Thus a reference quantity is, in effect, sent with the measuring quantities. The relative phases, and hence the character of the modulation, are changed due to dispersion in the cislunar medium. For example, a direct measure of the integrated ion density would be the value of the modulating frequency at which an amplitude-modulated signal is changed to a frequency-modulated echo. In another method an h-f and a vhf transmitter are modulated by the same audio signal. The difference of the audio phase in the simultaneously-recorded echoes should be a measure of the integrated ion density.

Doppler frequency-shifts and the earth's magnetic field cause only slight difficulties. However path splitting, ion "blobs", and lunar surface irregularities possibly could effect the waves more than does the integrated ion density. The measurements would then yield information about the temporal and spatial variations of the cislunar medium.

TABLE OF CONTENTS

LIST OF ILLUSTRATIONS

I. INTRODUCTION

During the past few years our ability to measure the ionized regions surrounding the earth has been extended to include greater and greater heights. The ionosphere up to the F-layer maximum (at about 1.05 earth radii) has been probed with radio sounders for many years. Polarization measurements on vhf radar echoes from the moon,[1,2] incoherent scatter measurements,[3,4] and measurements made with the aid of vertically-fired rockets and artificial earth satellites,[5,6] are being analyzed to find the ion density in the top side of the F-layer (up to about 1.15 earth radii). Very-low-frequency signals, originating from lightning and propagating along the earth's magnetic field lines, are being used to obtain preliminary estimates of the ion density out to about 5 earth radii.[7,8] Soon space vehicles to the moon and beyond will be instrumented to measure characteristics of the outer ionosphere and the interplanetary medium. Studies of the zodiacal light also yield some preliminary information about the interplanetary medium.[9]

In this paper are presented several presently-feasible techniques for using h-f radar echoes from the moon in an attempt to determine the integrated ion density between the earth and the moon (at about 60 earth radii). If the total density can be measured, the ionospheric component can be determined separately and subtracted, leaving a measure of the density of the interplanetary medium near the earth's orbit. The same techniques might be applied in the future to more distant h-f radar targets (Venus, Mars, and the sun) for measuring other parts of the interplanetary medium. In addition, signals from space vehicles might be used to measure integrated densities through application of the techniques considered here.

After presenting the various possible moon-echo techniques for determining cislunar ion densities, their limitations and relative feasibilities are discussed. It is shown that under certain conditions, instabilities of the medium or target may make a density measurement difficult or impossible. In a companion paper some preliminary experimental results are described.[10]

II. THE PULSE AND MODULATED C-W TECHNIQUES

While the cislunar medium is undoubtedly very tenuous, it may have a measurable effect on radio waves of relatively low radio frequency which have traveled the great distance to the moon and back. In particular, the phase and group velocities of the radio waves differ from their free-space value c; $v_{\emptyset} = c(1 - f_o^2/f^2)^{-1/2}$ and $v_g = c(1 - f_o^2/f^2)^{1/2}$, where $v_{\emptyset}$ and v_g are the phase and group velocities, f is the radio frequency, and f_o is the plasma frequency. The plasma frequency is related to the electron density through $f_o^2 = c^2 r_e N/\pi$, where r_e is the classical electron radius, 2.8178×10^{-15} m, and N is the number of electrons per cubic meter. The dimensional constant $c^2 r_e/\pi$ is about 80.6 m^3/t^2. (Rationalized MKS units will be used throughout this paper.)

The effects of the ionized medium on phase and group velocities lead to several pulse and modulated continuous-wave techniques for measuring the integrated ion density between the earth and the moon. Certain assumptions and simplifications are made in the initial derivations (such as neglecting magnetic fields, doppler shifts, collisions, and path deviations); the more important of these are discussed in section III.

A. PULSE DELAY

The propagation time T of a short pulse of radio energy, normalized by the free-space delay T_o, is

$$\frac{T}{T_o} = \frac{1}{T_o} \int \frac{ds}{v_g} \cong 1 + \frac{\overline{f_o^2}}{2f^2}$$

where $T_o = S/c$, $S = \int ds$ (the total length of path followed by the pulse), $\overline{f_o^2} = \int f_o^2 ds/S$ (the mean square of the plasma frequency along the path), and it is assumed that $f_o^2 \ll f^2$ at all points along the path. For radar targets $S = 2R$, where R is the range. If the range or free-space delay T_o is known and T is measured at known

f, $\overline{f_o^2}$ and hence the average electron density ($\overline{N} = \int Nds/S = \pi\overline{f_o^2}/c^2 r_e = \overline{f_o^2}/80.6$) can be determined from

$$\frac{T - T_o}{T_o} = \frac{\overline{f_o^2}}{2f^2} \tag{1}$$

Since T will be only slightly greater than T_o, T must be measured with great accuracy and T_o must be known very accurately for a determination of $\overline{N}$. For example, $T - T_o$ is about 250 μsec at $f = 20$ Mc and 10 μsec at $f = 100$ Mc for $\overline{N} = 10^9/m^3$ and $T_o = 2.5$ sec. The pulse delay method is illustrated in Fig. 1.

B. PULSE DELAY DIFFERENCE

By measuring the difference in propagation times ($T_1 - T_2$) of pulses at two frequencies f_1 and $f_2 > f_1$, target range or T_o need no longer be known to great precision. From the formulas above it follows that

$$\frac{T_1 - T_2}{T_o} = \frac{\overline{f_o^2}}{2}\left(\frac{1}{f_1^2} - \frac{1}{f_2^2}\right) \tag{2}$$

The pulse delay difference method is illustrated in Fig. 2. For $f_1 = 20$ Mc, $f_2 = 100$ Mc, $T_o = 2.5$ sec, and $\overline{N} = 10^9/m^3$, $T_1 - T_2$ is approximately 250 μsec. Note that if f in (1) or f_1 in (2) is in the upper vhf band or higher, the extra delay due to a cislunar ion density of $10^9/m^3$ becomes a few microseconds or less. Because of lunar surface irregularities, the range stability of the echoes is probably not sufficient for measurements of this precision. This is the reason for specifying h-f (3 - 30 Mc) moon echoes for the density determinations.

C. PULSE DISPERSION

The various frequencies which make up a pulse of r-f energy are

affected differentially by the dispersive medium. The pulse not only is retarded, but also its shape is altered. Thus it might be feasible to find $\overline{N}$ from the echo pulse shape, as compared with the transmitted pulse shape, without attempting to measure the extra delay. A rectangular r-f pulse of width τ has a broad frequency spectrum which can be characterized by the band from $f - 1/2\tau$ to $f + 1/2\tau$. The difference between the group delays at these two frequencies will be used here as a rough measure of the increase in pulse width ΔT (see also reference 11). Thus from (2), for $1/2\tau \ll f$,

$$\frac{\Delta T}{T_o} = \frac{\overline{f_o^2}}{f^3 \tau} = \frac{\overline{f_o^2} f_m}{f^3} \qquad (3)$$

The pulse dispersion method is illustrated in Fig. 3. For the last part of (3), τ was set equal to $1/f_m$, where f_m is a measure of the bandwidth of the transmitted pulse. At $f = 20$ Mc, $T_o = 2.5$ sec, and $\overline{N} = 10^9/m^3$, a transmitted 5 μsec pulse would be doubled in width by dispersion, while a 1 μsec pulse would be stretched out 25 μsec.

D. MODULATED C-W PHASE

It is well known that c-w (or long pulse) measurements can be made to yield essentially the same information as short pulse measurements for the same expenditure of radio energy. Often equipment consider- ations make it possible to produce more continuous r-f power than average pulse power, so it may be necessary, or at least advantageous, to use a c-w measuring technique. The problem of determining integrated ion density differs from the standard radar problem in that relative rather than absolute range information is desired. This leads to simplifications in both the theory and the experimental methods. The three c-w techniques which are equivalent to the above three pulse methods are presented below.

The number of r-f phase cycles θ_f along a path is given by $\theta_f = f \int ds/v_\phi$. Thus the ratio of θ_f to the number of r-f cycles in elapsed time T_o ($fT_o = \theta_{fo}$) is

$$\frac{\theta_f}{\theta_{fo}} = 1 - \frac{\overline{f_o^2}}{2f^2}$$

But of course it is not possible to measure θ_f directly. Its value in principle could be found within an unknown additive integer (on the order of tens of millions for h-f moon echoes) given a precise value of θ_{fo}. Since one integer in θ_f corresponds to a small fraction of a microsecond $(1/f)$ in time, while the expected effect of the medium is measured in hundreds of microseconds, neither could we hope to attain nor do we need (for reasonable accuracy) the ability to measure θ_f.

The basic period of uncertainty can be more closely related to the expected effect of the medium by sending two r-f waves at frequencies of f and $f + f_m$ $(f_m \ll f)$, with the difference frequency controlled from a stable modulating source at frequency f_m. Then the ratio of the number of modulation phase wavelengths θ_m along the path to the number of modulation cycles in elapsed time T_o $(f_m T_o = \theta_{mo})$ is

$$\frac{\theta_m}{\theta_{mo}} = 1 + \frac{\overline{f_o^2}}{2f^2}$$

so that

$$\frac{\theta_m - \theta_{mo}}{\theta_{mo}} = \frac{\overline{f_o^2}}{2f^2} \qquad (4)$$

where (4), of course, looks very much like (1). With proper choice of f_m and very accurate knowledge of θ_{mo}, the number of modulation cycles between θ_m and θ_{mo} can be small enough so that there are no integer uncertainties, yet still large enough to provide a good measure of $\overline{N}$. For example, if θ_{mo} is computed to be 2,491.21 ($f_m = 10^3$ cps and $T_o = 2.49121$ sec), and the echo phase is measured to be 0.46 cycles behind the f_m reference voltage, then $\theta_m - \theta_{mo}$ is an integer plus 0.25.

It follows that the integer would be zero if $\bar{N} = 10^9/m^3$, one if $\bar{N} = 5 \times 10^9/m^3$, etc. Thus by referring to past experience or by using other modulating frequencies, it may be possible to remove the integer uncertainty, though it would still be necessary to have precise information on the value of θ_{mo}. The modulated c-w phase method is illustrated in Fig. 4.

E. MODULATED C-W PHASE DIFFERENCE

As before, the requirement for precise target range information can be obviated by making the basic range measurement at two frequencies. If simultaneous modulated waveforms at f_1 and f_2 (i.e., f_1 and $f_1 + f_m$ for one and f_2 and $f_2 + f_m$ for the other waveform) are employed, the mean square plasma frequency can be found from

$$\frac{\theta_{m1} - \theta_{m2}}{\theta_{mo}} = \frac{\overline{f_o^2}}{2} \left(\frac{1}{f_1^2} - \frac{1}{f_2^2} \right) \tag{5}$$

The modulated c-w phase difference method is illustrated in Fig. 5. In (5), $\theta_{m1} - \theta_{m2}$ is the number-of-cycles lead of the modulation envelope in the echo at f_1 as compared with that at f_2, $f_2 > f_1$. It is evident that (5) is the c-w equivalent of the pulse delay difference method of (2). From experience, f_m can be chosen so that the phase lead is less than one cycle, or so that the number of integer cycles is known. Alternatively, two values of f_m could be used to resolve the integer ambiguity. For $f_1 = 20$ Mc, $f_2 = 100$ Mc, $f_m = 1000$ cpsec, $\theta_{mo} = 2500$ cycles, and $\bar{N} = 10^9/m^3$, $(\theta_{m1} - \theta_{m2})$ is approximately 0.25 cycles or $90°$. Note that it is no longer necessary for the source of f_m to be very stable.

F. MODULATED C-W DISPERSION

From the parallelism that has been established between the pulse and modulated c-w measuring techniques, it should be evident that we can write equation (6) from (3) before discussing its meaning. Thus

$$\frac{\Delta\theta_m}{\theta_{mo}} = \frac{\overline{f_o^2} f_m}{f^3} \tag{6}$$

where (6) also can be derived from (5) by considering f_1 and f_2 to be adjacent frequencies separated by f_m; i.e., $f_1 + f_m = f_2 = f$, $f_1 = f - f_m$, $f_2 + f_m = f + f_m$.

The quantity $\Delta\theta_m$ is the number-of-cycles lead of the modulation envelope of $f - f_m$ and f as compared with the modulation envelope of f and $f + f_m$. Alternatively, $\Delta\theta_m$ is the second difference (the difference of the two differences) of the r-f phases at $f - f_m$, f, and $f + f_m$. For example, if $f = 20$ Mc, $\theta_{fo} = 5 \times 10^7$ cycles, $f_m = 100$ kc, $\theta_{mo} = 2.5 \times 10^5$ cycles, and $\overline{N} = 10^9/m^3$, $\Delta\theta_m$ is 0.25 cycles or $90°$. For these precise values, $\theta_{f - f_m}$, θ_f, and $\theta_{f + f_m}$ would be $49,744,974.8744$, $49,995,000.000$, and $50,245,024.8756$ cycles respectively, so that θ_{m1} and θ_{m2} would be $250,025.1256$ and $250,024.8756$ cycles respectively, and $\Delta\theta_m$ would be 0.2500 cycles. The beauty of the dispersion method lies in the fact that no attempt is made to determine the θ_f's or θ_m's, but $\Delta\theta_m$, which is proportional to $\overline{N}$, is obtained directly from the relative phases in the echo. As with the phase difference method, it is not even necessary for f_m to be highly stable.

The modulated c-w dispersion method is illustrated in Fig. 6. This method would be particularly convenient in practice since the three transmitted frequencies with equal spacings can be supplied using sine-wave amplitude or angle (frequency or phase) modulation at f_m on a carrier at f. Dispersion during propagation changes the relative phases and hence the character of the modulation. In terms of either amplitude or angle modulation, $\Delta\theta_m$ is twice the increase in phase lead of the carrier relative to the phase of the sum of the two sidebands. For example, if f_m is chosen so that $\Delta\theta_m = 0.5$ cycles or $180°$, an amplitude-modulated transmitted wave will be received as an angle-modulated echo, and vice versa. This value of f_m is then a direct measure of the integrated ion density between the earth and the moon.

III. SOME LIMITATIONS

The relative feasibility of the six density-measuring techniques will be determined by both practical and environmental limitations. The practical limitations include transmitter power capabilities and

lunar range knowledge. Doppler shift, polarization rotation, medium stability and uniformity, and lunar roughness and libration will be considered with regard to what limitations they may impose. These factors were omitted in the above discussion for the sake of clarity in the general presentation.

A. TRANSMITTER POWER

With a very high gain antenna and a source of several kilowatts of c-w power at h-f, reasonably strong moon echoes can be obtained using a receiver bandwidth on the order of 100 cps. Thus several megawatts of peak power would be required for pulse measurements using pulse lengths on the order of ten microseconds. In view of the present relative availability of h-f transmitters suitable for pulse or c-w moon-echo measurements, it appears that the c-w techniques described above must be considered to be more feasible than the corresponding pulse methods.

B. TRUE RANGE TO THE MOON

For the pulse delay and modulated c-w phase measurements, the true lunar range must be known with great precision. If the actual time delay to the leading edge of the echo were measured with perfect accuracy, T_o ($T_o = \theta_{mo}/f_m = 2R/c$) would have to be known to one part in 10^5 (25 μsec in 2.5 sec) in order to determine $\overline{N}$ to an accuracy of 10% at 20 Mc, assuming $\overline{N}$ is near $10^9/m^3$. At 100 Mc, for the same T_o accuracy and true value of $\overline{N}$, the measurement for $\overline{N}$ might come out anywhere between minus $1.5 \times 10^9/m^3$ to plus $3.5 \times 10^9/m^3$. Even at 20 Mc a measurement of density based on radar and true range difference would be unreliable if $\overline{N}$ were only $10^8/m^3$.

In moon echo measurements using 2 μsec pulses at 3000 Mc, where the true and radar ranges should be sensibly equal, Yaplee, Bruton, Craig, and Roman[12] found differences with a mean of 45 μsec between computed and measured range. By using a different value for the earth's equatorial radius they were able to reduce the mean residual to near zero, but fluctuations up to 10 μsec remained. Until more work is done at various positions on the earth with precision vhf and uhf radars, it appears that the various uncertainties (in the earth-center to moon-center distance, in the size and shape of the earth and moon, and in the pre-

cise location of the radar on the earth and the reflecting point on the
moon) are such that the 25 μsec uncertainty figure used above is real-
istic. Thus, unless the cislunar ion density is considerably greater
than the assumed value near $10^9/m^3$, the pulse delay and modulated c-w
phase techniques must be considered less feasible than the difference or
dispersion methods.

C. DOPPLER

Due to both the rotation of the earth and the motion of the moon
in its non-circular orbit, the radio frequency of the echo will differ
from the transmitted frequency. The echo frequency is approximately
$(1 - 2v/c)$ times the transmitted frequency, where v is the radial
velocity of the reflecting point relative to the position of the radar.
The maximum fractional change in frequency due to the Doppler effect is
less than 10^{-5} for moon echoes; i.e. $2v/c < 10^{-5}$.

The Doppler change in frequency affects the precise interpretation
of h-f moon echoes for the determination of ion densities between the
earth and the moon. However in each of the pulse techniques, the
fractional effect of Doppler on the determination of $\overline{N}$ is negligible,
being only on the order of $2v/c$. The same is true of each of the c-w
techniques, although it should be mentioned that the approximate
Doppler frequency may need to be known in order to place and keep the
echoes in the center of the pass band of very-narrow-band receivers.

D. MAGNETIC FIELDS

With the simultaneous presence of ions and an ordered magnetic
field, the propagating medium is doubly refractive as well as dispersive.
For frequencies well above the plasma frequency and for all directions
of propagation which are more than a few degrees from the normal to the
magnetic field, a linearly-polarized transmitted wave can be considered
as consisting of two equal-strength, oppositely-rotating, circularly-
polarized waves having different phase velocities. Thus the sum of
these two waves is a linearly-polarized wave whose spatial direction of
polarization changes with position along the direction of propagation
(Faraday rotation). For a transmitted circularly-polarized wave, only
one of the two normal modes is excited.

To include the magnetic field under the above condition on direction of propagation, the index of refraction is changed from $\left[1 - f_o^2/f^2\right]^{1/2}$ to $\left[1 - (f_o^2/f^2) / (1 \pm f_L/f)\right]^{1/2}$, with a corresponding change in the expression for phase velocity. The effect on group velocity is not so simple, but for frequencies well above the plasma frequency and for $f_L \ll f$, the approximate expressions become

$$v_\emptyset = c \left[1 + (f_o^2/2f^2) / (1 \pm f_L/f)\right]$$

and

$$v_g = c \left[1 - (f_o^2/2f^2) / (1 \pm f_L/f)^2\right]$$

In the above, f_L is the longitudinal gyro frequency, $f_L = (2r_e/e)H \cos \theta$, where e is the magnitude of the charge on an electron, 1.602×10^{-19} coulombs, H is the magnetic field intensity in amperes per meter (about 25 and 50 a/m near the surface of the earth, at the equator and pole, respectively), and θ is the acute angle between the direction of propagation and the magnetic field. The upper sign corresponds to the so-called ordinary wave and the lower sign to the extraordinary wave.

With the same approximations as have been used above, the six moon-echo formulas for determining cislunar ion densities can be altered to include the earth's magnetic field by changing f_o^2 to $f_o^2(1 \mp 2f_L/f)$ in (1) and (4), to $f_o^2 \left\{1 \mp (2f_L/f_1) \left[1 + f_1^2/f_2(f_1 + f_2)\right]\right\}$ in (2) and (5), and to $f_o^2(1 \mp 3f_L/f)$ in (3) and (6).

If a circularly-polarized wave were transmitted or received, only the appropriate one of the double signs in the above expressions would be used; the upper sign for ordinary and the lower sign for extra-ordinary waves. For a linearly-polarized wave the two normal modes are equally excited so that the total effect of the field is half the sum of the effects computed for the upper and lower signs. But this brings us back to the no-field case as far as computations of density and echo phase are concerned. Of course, there are still magnetic field effects to second and higher orders of f_L/f, but they can be neglected since f_L^2/f^2 is less than 0.01 for frequencies above 20 Mc.

- 10 -

As indicated above, measurements with a single circularly-polarized wave would yield a complex average involving gyro and radio frequencies as well as plasma frequency. To find $\overline{N}$ from this complex average it would be necessary to estimate (or measure by another experiment) the value of f_L and f_o^2 through the ionosphere, out to the radius at which the correction terms are negligible. This estimation or added experiment would be very critical if the integrated ion density through the ionosphere greatly exceeds the integrated ion density through the cislunar medium beyond about 1000 km. However, if the ionospheric component is small relative to the rest of the path, no magnetic field correction need be made. If there are, on the average, 10^9 electrons per cubic meter beyond the ionosphere, the ionospheric contribution will vary from about 0.1 to 0.5 of the total integrated density.

Until more is known of the relative effects of the ionosphere and the regions beyond, it appears important to avoid the use of a single circularly-polarized antenna for either transmitting or receiving. The best arrangement would be to transmit a linearly-polarized wave, thereby equally exciting the two magneto-ionic modes, and to receive in an orthogonal pair. If reception were in a single, fixed, linearly-polarized antenna instead of an orthogonal pair, there would be selective fading due to Faraday rotation and a consequent change of uncertainty in phase measurements from n cycles to 2n half-cycles. While these effects would be troublesome, they could be coped with more easily than the difficulty involved in the use of only one of the two ionospheric modes. Thus, while transmitting linear polarization and receiving an orthogonal pair would be the best arrangement, the next best procedure would be to transmit linear and receive linear polarization.

It might be feasible to combine the dispersion measurement with a measurement of Faraday rotation to obtain, in one experiment, both the ionospheric and the cislunar integrated ion densities. As before there is a correspondence between the pulse and c-w methods, but the selective fading in the echo pulse due to Faraday rotation would be difficult to interpret. (This fading would also affect adversely the pulse dispersion method of measuring the total density.) Thus we will consider here only

the c-w or long-pulse measurements.

The number of r-f cycles along the two magneto-ionic modes is given simply by $\theta_{f1} = f \int ds/v_{\emptyset 1}$ and $\theta_{f2} = f \int ds/v_{\emptyset 2}$, where the subscripts correspond to the ordinary (1) and extraordinary (2) waves. Assuming equal excitation of these two modes, the number of spatial revolutions of a linearly-polarized wave Ω is given by

$$\Omega \;=\; \frac{\theta_{f1} - \theta_{f2}}{2}$$

so that, for $f_L/f \ll 1$,

$$\frac{\Omega}{\theta_{f0}} \;=\; \frac{\overline{f_0^2} \, f_L}{2f^3} \tag{7}$$

For example, if $f_L = 1.6$ Mc and $\overline{f_0^2} = 10^{13}$ for the first 1500 km of the ionosphere ($T_0 = 10^{-2}$ sec), the plane of polarization on a round trip through the ionosphere is changed by 200 revolutions at 20 Mc and 8 revolutions at 100 Mc.

The difference in the number of cycles of rotation $\Delta\Omega$ for two closely-spaced frequencies f and $f + f_m$ becomes, from (7)

$$\frac{\Delta\Omega}{\theta_{mo}} \;=\; \frac{\overline{f_0^2} \, f_L}{f^3} \tag{8}$$

With the above assumed parameters and a carrier frequency of 20 Mc, $\Delta\Omega = 0.02$ revolutions for $f_m = 1$ kc and $\Delta\Omega = 2$ revolutions for $f_m = 100$ kc. Thus if the polarization of the echo were measured, it appears that modulating frequencies could be chosen so as to combine either the modulated c-w phase difference or modulated c-w dispersion measurement of total cislunar ion density with the Faraday rotation method of measuring the integrated ion density through the earth's ionosphere.

E. MEDIUM STABILITY AND LUNAR ROUGHNESS

In the above derivations it has been assumed that all of the radio energy travels along the same path. But different frequencies follow different paths because of gradients of ionization associated with layers and "blobs" in the ionosphere and in the rest of the cislunar medium. The earth's magnetic field causes additional path splitting. The phase center of the moon may be a function of frequency and polarization. Even at a single frequency, the phase front of the wave is not plane. All of these effects can be expected to change with time. These temporal and spatial variations of the moon and the cislunar medium may affect adversely the various methods of measuring integrated ion density.

From a preliminary theoretical investigation of path splitting, it appears convenient to consider two effects separately. The waves following different paths will encounter different ion densities and will travel different total distances. A combined effect of these differences will exceed the effects given in (1) through (6) unless

$$\frac{\Delta(\overline{Sf_o^2})}{\overline{Sf_o^2}} < \frac{f_m}{f} \tag{8}$$

where $S = \int ds$, $\overline{Sf_o^2} = \int f_o^2 ds$, and the Δ is used to denote the difference of the following quantity at f and $f + f_m$. In addition, the effect of the difference in path length alone will exceed the density effects given in (1) through (6) unless

$$\frac{\Delta S}{S} < \frac{\overline{f_o^2} f_m}{2f^3} \tag{9}$$

Both (8) and (9) must be satisfied if the measured pulse and c-w effects are to be interpreted as being due primarily to the integrated cislunar ion density.

A simple theoretical model of an ionized region with gradients has

been investigated, and it is found that $\triangle S/S$ is on the order of $\overline{\triangle f_o^2} \, f_m/f^3$. Thus if this model is at all representative of the ionosphere and the rest of the cislunar medium, (9) will be satisfied. Since the right side of (9) is only $\overline{f_o^2}/2f^2$ of f_m/f, we now need to consider only $\overline{\triangle f_o^2}/f_o^2$ in (8). For example this ratio will be less than f_m/f, with $\overline{N} = 10^9/m^3$ and f = 20 Mc, if the average electron density along one path differs from that along another by less than $5 \times 10^4/m^3$ for f_m = 1 kc or $5 \times 10^6/m^3$ for f_m = 100 kc. If these conditions are not satisfied the ion densities deduced from the h-f moon echoes will not be accurate. The temporal behavior of the echoes should provide a clue as to whether the effect of the integrated ion density or of path splitting is being measured.

The phase center of the moon may be different at different frequencies. From (9), the difference in phase center must be less than 4 m at f_m = 1 kc or 400 m at f_m = 100 kc (for $S = 2R = 8 \times 10^8$ m, $\overline{N} = 10^9/m^3$, and f = 20 Mc) if this effect is to be less than that of the integrated ion density. (This appears to be a very stringent requirement until we consider that it corresponds to a 250 μsec pulse-delay error at 20 Mc.) As the moon changes aspect relative to the radar due to lunar libration, the phase center and the difference in phase center would be expected to change. Again the temporal variations of the echo characteristics may help determine the relative effects of the lunar surface irregularities and the integrated density of the cislunar medium.

It is interesting to note that a very rough moon, which would cause much spreading in a short pulse, could have a stable phase center for c-w. This appears to be one aspect of the moon-echo measurements in which there is no simple correspondence between pulse and several-frequency c-w techniques. If c-w amplitude and phase measurements were made at all of the frequencies included in the short pulses, the equivalent information would, of course, be obtained.

IV. CONCLUSIONS

After considering the limitations of the various methods for

measuring the cislunar ion density, it now appears that either the modulated c-w phase difference (5) or the modulated c-w dispersion (6) method will prove to be best. In both these techniques there are no stringent equipment stability problems nor is there need for precise knowledge of the lunar range. The reference quantities are sent along with the measuring quantities. The required modulation is very simple. The effect of the earth's magnetic field can be either avoided or used to obtain a simultaneous measurement of the integrated ionospheric ion density. Doppler changes in frequency are not important. The c-w dispersion method has the added advantage of only requiring one trans- mitter. However, this method requires that there be a stable phase response of medium and moon over a wider band of frequencies in the h-f band than is required for the modulated c-w phase difference method. It may be that the phase difference method will prove better if the time and space variations of the cislunar medium or of the lunar surface are troublesome. On the other hand, simultaneous phase measurements at a large number of frequencies may be needed to average out the effects of these variations. One or another of the pulse methods would then be favored over the c-w methods since they automatically average over a band of frequencies.

Even if the irregularities of the medium and moon cause large phase or delay effects, it still may be possible to obtain a good measure of the integrated ion density by time averaging. The effects of the medium and moon irregularities should change with time, and they probably are equally likely to increase or decrease the delay or phase. The ion density effects, on the other hand, should be relatively stable and will always change the delay or phase in the same direction.

It is probable that the "roughness" of the cislunar medium will increase in importance as the carrier frequency is lowered, while the roughness of the lunar surface will increase in importance (relative to the integrated density effects) as the frequency is raised. It is even possible that these effects will leave no room in the middle for a successful measurement of the cislunar density. Early studies of h-f moon echoes led to the conclusion that the moon was a very rough reflector[13], while later studies at vhf and uhf show that it is relatively

smooth[12] (though still too rough for a density measurement at these frequencies). It now appears that the moon is relatively smooth, but at h-f the cislunar medium may be rough.

If one of the above techniques proves to be a good way for measuring the cislunar ion density, it will be very interesting to find how the density of this medium varies with various solar and terrestrial phenomena. The measurement techniques described here may also be applicable in the near future to other radar and space probe investigations of the density of the interplanetary medium. Included in such experiments would be radar echoes from Venus, Mars, the sun, and artificial satellites, and radio signals from satellites, lunar probes, and planetary probes.

BIBLIOGRAPHY

1. J. V. Evans, "The electron content of the ionosphere," Jour. Atmos. Terr. Phys., vol. 11, pp. 259-271; 1957.

2. S. J. Bauer and F. B. Daniels, "Ionospheric parameters deduced from the Faraday rotation of lunar radio reflections," Jour. Geophys. Res., vol. 63, pp. 439-442; June, 1958.

3. W. E. Gordon, "Incoherent scattering of radio waves by free electrons with applications to space exploration by radar," Proc. IRE, vol. 46, pp. 1824-1829; Nov., 1958.

4. K. L. Bowles, "Observation of vertical-incidence scatter from the ionosphere at 41 Mc/sec," Phys. Rev., vol. 1, Dec. 15, 1958.

5. H. Friedman, "Rocket observations of the ionosphere," Proc. IRE, vol. 47, pp. 272-280; Feb., 1959.

6. O. K. Garriott, "Ionospheric electron content determined from satellite observations," paper presented at Spring URSI Meeting, Washington, D. C., May 4-7, 1959.

7. L. R. O. Storey, "An investigation of whistling atmospherics," Phil. Trans. Roy. Soc. A., vol. 246, pp. 113-141; July, 1953.

8. R. A. Helliwell and M. G. Morgan, "Atmospheric whistlers," Proc. IRE, vol. 47, pp. 200-208; Feb., 1959.

9. D. E. Blackwell, "A study of the outer corona from a high altitude aircraft at the eclipse of 1954 June 30. I. Observational data," Mon. Not. Roy. Astron. Soc., vol. 115, pp. 629-649; 1955.

10. P. B. Gallagher, R. C. Barthle, and V. R. Eshleman, "Preliminary results of radar studies of the cislunar medium," Scientific Report No. 6, contract AF19(604)-2193, Stanford University (in preparation).

11. V. A. Counter and E. P. Riedel, "Calculations of ground-space propagation effects," Lockheed Aircraft Corporation Report LMSD-2461; May, 1958.

12. B. S. Yaplee, R. H. Bruton, K. J. Craig, and N. G. Roman, "Radar echoes from the moon at a wavelength of 10 cm," Proc. IRE, vol. 46, pp. 293-297; Jan., 1958.

13. F. J. Kerr and C. A. Shain, "Moon echoes and transmission through the ionosphere," Proc. IRE, vol. 39, pp. 230-242; March, 1951.

$$\frac{T - T_0}{T_0} = \frac{\overline{f_0^2}}{2f^2}$$

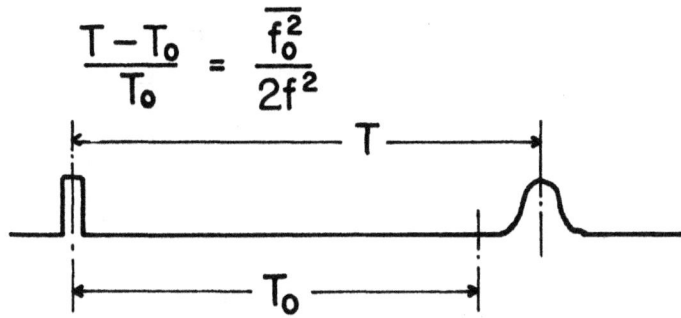

FIG. 1.--PULSE DELAY.

$$\frac{T_1 - T_2}{T_0} = \frac{\overline{f_0^2}}{2}\left(\frac{1}{f_1^2} - \frac{1}{f_2^2}\right)$$

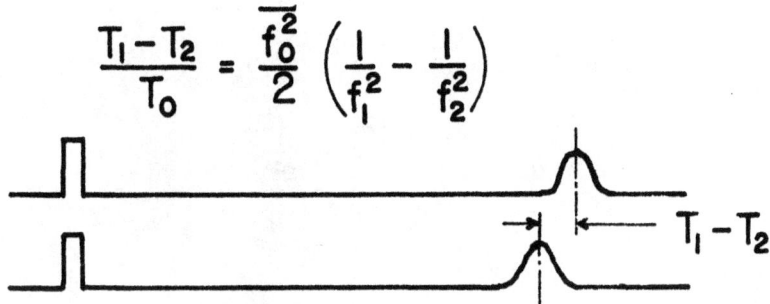

FIG. 2.--PULSE DELAY DIFFERENCE.

$$\frac{\Delta T}{T_0} = \frac{\overline{f_0^2}\, f_m}{f^3}$$

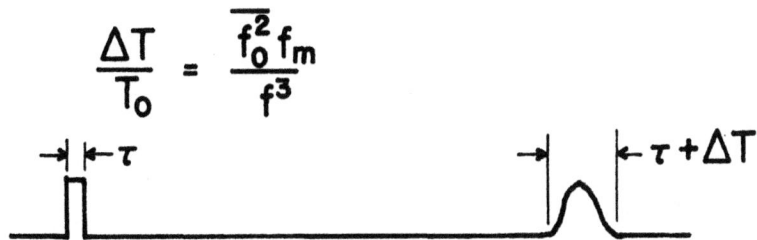

FIG. 3.--PULSE DISPERSION.

$$\frac{\theta_m - \theta_{mo}}{\theta_{mo}} = \frac{\overline{fo^2}}{2f^2}$$

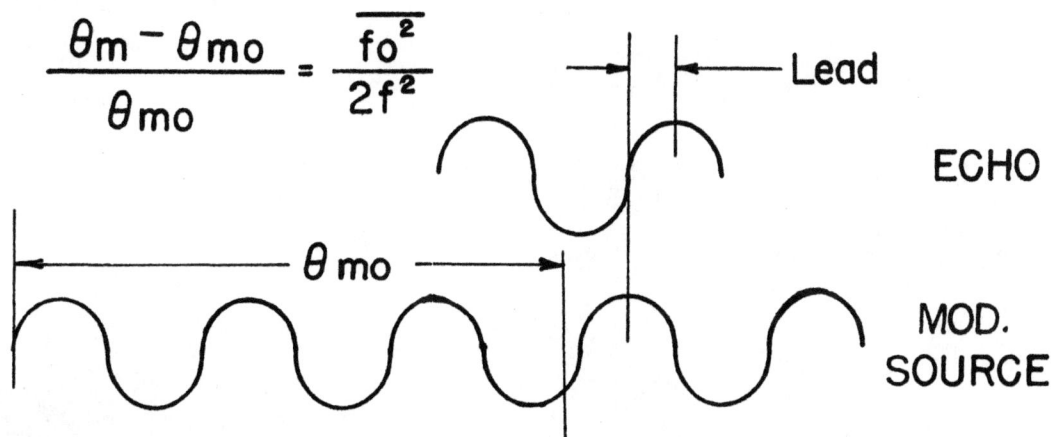

FIG. 4.--MODULATED C-W PHASE.

$$\frac{\theta_{m1} - \theta_{m2}}{\theta_{mo}} = \overline{\frac{f_o^2}{2}} \left(\frac{1}{f_1^2} - \frac{1}{f_2^2} \right)$$

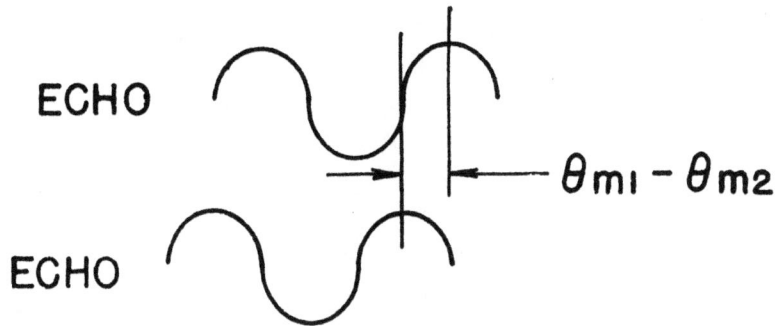

ECHO

ECHO

$\theta_{m1} - \theta_{m2}$

FIG. 5.--MODULATED C-W PHASE DIFFERENCE.

$$\frac{\triangle \theta_m}{\theta_{mo}} = \overline{\frac{f_o^2 f_m}{f^3}}$$

$$\triangle \theta_m = (\theta_L - \theta_c) - (\theta_c - \theta_u) \quad \text{3 ECHOES}$$

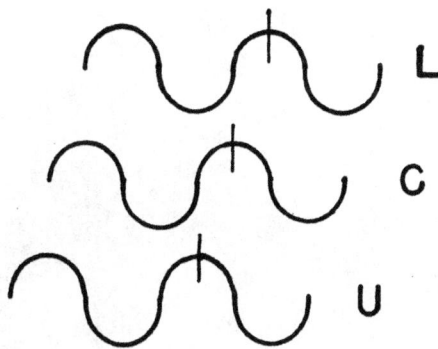

L

C

U

FIG. 6.--MODULATED C-W DISPERSION.

Distribution List
for Scientific Report No. 5
Contract AF 19(604)-2193

No.	Code	Organization
1	AF 5	Air Force Missile Test Center Patrick Air Force Base, Florida Attn: MU-411, Technical Library
1	AF 18	Air University Library Maxwell Air Force Base, Alabama
1	AF 41	Wright Air Development Center (WCOSR) Foreign Release Branch Wright-Patterson Air Force Base, Ohio
1	AF 43	Wright Air Development Center (WCLJA-2) Aeronautical Research Lab. Research Division Wright-Patterson Air Force Base, Ohio
1	AF 65	Wright Air Development Center Wright-Patterson Air Force Base, Ohio Attn: WCLNQ-4, Mr. Draganjac
1	AF 68	Wright Air Development Center Wright-Patterson Air Force Base, Ohio Attn: WCLRE-5, Mr. P. Springer
2	AF 91	Air Force Office of Scientific Research Washington 25, D. C. Attn: SRY, Mr. Wm. J. Otting
1	AF 121	Rome Air Development Command Griffiss Air Force Base, New York Attn: RCUO, Mr. Charles A. Strom, Jr.
1	AF 123	Wright Air Development Center Wright-Patterson Air Force Base, Ohio Attn: WCLNE-4, Mr. Klaus Otten
1	AF 124	Rome Air Development Center Griffiss Air Force Base, New York Attn: RCSSTL-1
1	AF 157	Wright Air Development Center Wright-Patterson Air Force Base, Ohio Attn: WCLNO, Mr. R. E. Kester
1	AF 166	Headquarters, USAF (AFOAC-S/E) Washington 25, D. C. Attn: Communications-Electronics Directorate

No.	Code	Organization
1	AF 178	Air Research and Development Command Andrews Air Force Base Washington 25, D. C. Attn: RDTC
1	AF 182	Rome Air Development Command Griffiss Air Force Base, New York Attn: Mr. T. G. Knight, RCUELL
1	AF 184	Headquarters, USAF, Washington 25, D. C. Attn: Lt. Col. W. Hodson, AFDRD
1	AF 191	TAC (Dir. of Telecommunications Systems) Langley Air Force Base, Virginia
1	Ar 5	US Army Signal Engineering Labs Technical Documents Center Evans Signal Laboratory Belmar, New Jersey
1	Ar 7	Office of Chief Signal Officer US Army Radio Frequency Eng. Office Attn: SIGFO-B4, Room BD 973, G. W. Haydon The Pentagon, Washington 25, D. C.
1	Ar 9	Department of the Army Office of the Chief Signal Officer Washington 25, D. C. Attn: OCSigO, Research and Dev. Division
2	Ar 28	Asst. Sec. of Defense for Research and Dev. Information Office Library Branch Pentagon Building, Washington 25, D. C.
1	Ar 32	Mr. Frederic H. Dickson, Chief US Army Signal Radio Propagation Agency Fort Monmouth, New Jersey
1	Ar 52	Department of the Army Evans Signal Laboratory Belmar, New Jersey Attn: Alexander N. Beichek
1	Ar 68	U.S. Army Signal Engineering Laboratories Fort Monmouth, New Jersey Attn: Mr. Robert Kulinyi, Long Range Radio Branch
10	G 2	ASTIA Arlington Hall Station Arlington 12, Virginia
2	G 6	Office of Technical Services Department of Commerce Washington 25, D. C. Attn: Tech. Reports Section

No.	Code	Organization
2	G 8	Library, Boulder Laboratories National Bureau of Standards Boulder, Colorado
1	G 18	Office of Science Office of Asst. Sec. of Defense, Res. and Eng. Washington 25, D. C.
1	G22	Federal Communications Commission Technical Research Division Washington 25, D. C. Attn: Mr. Harry Fine
1	G 26	Central Radio Propagation Laboratory National Bureau of Standards Boulder, Colorado Attn: K. A. Norton
1	G 46	Ionospheric Research Section Central Radio Propagation Laboratory National Bureau of Standards Boulder, Colorado Attn: R. C. Kirby, Chief
1	G 54	U. S. Department of Commerce National Bureau of Standards Central Radio Propagation Laboratory Boulder, Colorado Attn: Mr. G. R. Sugar, Ionospheric Res. Sec.
1	G 60	U.S. Advisory Group, US Embassy, The Hague Attn: Mr. Pierre Bartholeme, SADTC Department of State, Washington 25, D. C.
1	I 2	Stanford Research Institute Menlo Park, California Attn: Dr. J. T. Bolljahn, Engineering Res.
1	I 53	Hughes Aircraft Company Florence Ave at Teale Street Culver City, California Attn: Documents Section, Res. and Dev. Library
1	I 103	Stanford Research Institute Menlo Park, California Attn: Dr. A. M. Peterson
1	I 116	Melpar, Inc. 3000 Arlington Boulevard Falls Church, Virginia Attn: Mr. C. B. Raybuck, Chief Engineer

No.	Code	Organization
1	I 128	Raytheon Manufacturing Company Service Building, Wayland Lab. Wayland, Mass. Attn: D.A. Hedlund, Section Manager Communications Department
1	I 132	Bell Telephone Laboratories Murray Hill, New Jersey Attn: Mr. K. Bullington
1	I 133	Collins Radio Company 855-35th Street, N.E. Cedar Rapids, Iowa Attn: Irvin H. Gerks
1	I 198	RCA Laboratories David Sarnoff Research Center Princeton, New Jersey Attn: Mr. Richard Jenkins
1	I 225	Pickard and Burns, Inc. 240 Highland Avenue Needham 94, Mass. Attn: Dr. R. Woodward
1	L 226	Rand Corporation 1700 Main St. Santa Monica, California Attn: E. E. Reinhart
1	I 598	Lockheed Aircraft Corporation Electronics and Armaments Systems Div. Building 63, Plant A-1, Burbank, California Attn: Mr. Gene Garrison
1	I 636	General Electric Company Defense Electronics Division 735 State Street, Santa Barbara, Calif. Attn: Mr. Klaus G. Likehold, Manager-Library
1	I 639	Ferranti-Packard Electric Electronics Division Industry Street Toronto 15, Ontario, Canada Attn: Mr. G. W. L. Davis
1	I 640	RCA Laboratories Riverhead, Long Island, New York Attn: Mr. Robert Wagner
2	M 5	Air Force Cambridge Research Center Laurence G. Hanscom Field Bedford, Massachusetts Attn: CROTLR-2, P. Condon

No.	Code	Organization
	M 6	Air Force Cambridge Research Center Laurence G. Hanscom Field Bedford, Massachusetts Attn: CROTLS, J. Armstrong
1	M 28	Air Force Cambridge Research Center Laurence G. Hanscom Field Bedford, Massachusetts Attn: CRRST, Mr. Leon Ames
9	M 29	Air Force Cambridge Research Center Laurence G. Hanscom Field Bedford, Massachusetts Attn: CRRK, Mr. W. Griffin
2	N 1	Director, Avionics Division (AV) Bureau of Aeronautics Department of the Navy Washington 25, D. C.
1	N 14	Office of the Chief of Naval Operations Department of the Navy OP-583 Washington 25, D. C.
1	N 17	Bureau of Ships, Code S19 Department of the Navy Washington 25, D. C. Attn: Mr. R. S. Baldwin
1	N 27	Librarian U.S. Naval Postgraduate School Monterey, California
1	N 28	Air Force Development Field Representative Naval Research Laboratory Code 1072, Washington 25, D. C.
2	N 29	Director, U.S. Naval Research Lab. Washington 25, D. C. Attn: Code 2027
1	N 45	Commanding Officer and Director U.S. Navy Electronics Lab San Diego 52, California Attn: D.P. Heritage, Head, Radio Branch
1	N 48	Commanding Officer U. S. Naval Air Development Center Johnsville, Pennsylvania Attn: NADC Library

No.	Code	Organization
2	N 37	Chief of Naval Research Electronics Branch (Code 427) Department of the Navy Washington 25, D. C. Attn: Dr. Arnold Shostak
2	N 74	Chief, Bureau of Ships Department of the Navy Washington 25, D. C. Attn: 310
1	N 75	Chief, Bureau of Ordnance Department of the Navy Washington 25, D. C. Attn: Code Rep-a
1	N 85	Commanding Officer and Director U.S. Navy Electronics Laboratory (library) San Diego 52, California
1	U 14	University of Florida Electrical Engineering Department Gainesville, Florida Attn: Dr. George
2	U 23	Applied Physics Laboratory Johns Hopkins University 8621 Georgia Avenue Silver Springs, Maryland
1	U 32	Massachusetts Institute of Technology Research Laboratory of Electronics Room 20B-221 Cambridge 39, Massachusetts Attn: John H. Hewitt
1	U 37	The University of Michigan Engineering Research Institute Willow Run Laboratories Willow Run Airport Ypsilanti, Michigan Attn: Librarian
1	U 49	Stanford University Stanford, California Attn: Dr. Von R. Eshleman Radio Propagation Laboratory
1	U 63	Cornell University School of Electrical Engineering Ithaca, New York Attn: Dr. W. Gordon

No.	Code	Organization
1	U 64	University of Texas Electrical Eng. Research Lab. Box 8026 University Station Austin 12, Texas Attn: Prof. A. W. Straiton
1	U 69	Georgia Institute of Technology Engineering Experiment Station Atlanta, Georgia Attn: W. B. Wrigley, Head Communications Br.
1	U 83	Control Systems Laboratory University of Illinois Urbana, Illinois Attn: Dr. F. W. Loomis, Director
1	U 84	Stanford University Stanford, California Attn: Dr. Oswald G. Villard, Jr. Radio Propagation Laboratory
1	U 85	University of Tennessee Department of Electrical Engineering Knoxville, Tennessee Attn: Prof. F. V. Schultz
1	U 140	Massachusetts Institute of Technology Lincoln Laboratory P. O. Box 73 Lexington 73, Mass. Attn: Mr. J. H. Chisholm - C-357
1	U 243	Massachusetts Institute of Technology Lincoln Laboratory P. O. Box 73 Lexington 73, Mass. Attn: Dr. Morton Loewenthal
1	I 187	Electromagnetic Research Corporation 711 - 14th Street, N. W. Washington 5, D. C. Attn: Mr. Martin Katzin
1	I 638	AVCO Research Laboratory 2385 Revere Beach Parkway Everett 48, Mass. Attn: Technical Librarian
1	I 648	The Mitre Corporation 244 Wood Street Lexington 73, Mass. Attn: Mrs. Jean E. Claflin, Librarian